The 50 Hike

Written by B. Y. Wral
Illustrated by Curt Walstead

"Are you ready?" asked Ms. Cutter.

"Let's go on a 50 hike!" she said.
"Let's find 50 of one thing."

"Stop," said Ms. Cutter.
"Let's count.
How many flowers do you see?"

"I count 12," said Annie.

"You are right!" said Ms. Cutter.
"There are 12 flowers."

"Let's hike!
We must find 50 of one thing."

"Stop," said Ms. Cutter.
"Let's count.
How many ducks do you see?"

"I count 21," said Andy.

"You are right!" said Ms. Cutter.
"There are 21 ducks."

"Let's hike!
We must find 50 of one thing."

"Stop," said Ms. Cutter.
"Let's count the balloons."

Your Turn

How many balloons do you see?

"You are right," said Ms. Cutter.
"There are 34 balloons.
Let's hike.
We must find 50 of one thing!"

"Look," said Annie.
"I found some cookies.
There are 50 chocolate chips."

"We found 50 of one thing!" said Andy.

"Good job!" said Ms. Cutter.
"Let's eat!"